TESTIMONY ON ASSURED ACCESS TO SPACE

WEDNESDAY, JULY 16, 2014

U.S. SENATE,
SUBCOMMITTEE ON STRATEGIC FORCES,
COMMITTEE ON ARMED SERVICES;
AND COMMITTEE ON COMMERCE,
SCIENCE, AND TRANSPORTATION,
Washington, DC.

The committees met, pursuant to notice, at 9:29 a.m. in room SH–216, Hart Senate Office Building, Senator Bill Nelson (chairman of the subcommittee) presiding.

Committee members present: Senators Nelson, Udall, Donnelly, Kaine, King, McCain, Sessions, Wicker, Lee, and Cruz.

Committee staff members present: Peter K. Levine, staff director; and Leah C. Brewer, nominations and hearings clerk.

Majority staff member present: Jonathan S. Epstein, counsel.

Minority staff member present: Daniel A Lerner, professional staff member.

Staff assistants present: Lauren M. Gillis, Daniel J. Harder, and Brendan J. Sawyer.

Committee members' assistants present: Cathy Haverstock, assistant to Senator Nelson; Christopher R. Howard, assistant to Senator Udall; Rachel H Lipsey, assistant to Senator Donnelly; Nick Ikeda, assistant to Senator Hirono; Karen E. Courington, assistant to Senator Kaine; Stephen M. Smith, assistant ton Senator King; Jeremy H. Hayes, assistant to Senator McCain; Lenwood A. Landrum, assistant to Senator Sessions; Joseph G. Lai, assistant to Senator Wicker; Bradley L. Bowman, assistant to Senator Ayotte; and Peter H. Blair, assistant to Senator Lee.

OPENING STATEMENT OF SENATOR BILL NELSON

Senator NELSON. Good morning. As Senator Udall and Senators Sessions and Cruz arrive, I will recognize them. I want to get us going because we are facing a couple of votes this morning. So we are just going to have to play this by ear. We will try to keep the hearing going.

Forty five years ago it was today that Apollo 11 launched. Most everybody that is a certain age and older in this room will remember exactly where they were on that day. And 4 days later, of course, Armstrong and Collins became the first to set their foot on the moon.

And in the decades since, space technology has become vital to our Nation's security, economy, and standard of living. And so it is appropriate that we are holding this hearing to discuss reliable

(1)

domestic space access, and that is the bottom line of what we are trying to achieve is the goal of assured access to space by American vehicles for both unmanned and manned payloads.

Obviously, the tensions with Russia as a result of the Ukraine have forced us to rethink part of the relationships that have built up, and that is despite decades of cooperation, first with the Soviets. And, of course, just remember in the midst of the Cold War, an American spacecraft and a Soviet spacecraft rendezvoused and docked and the crews lived together for 9 days in space. And those crews are personal good friends, and the personal relationship, as exhibited by Tom Stafford and Alexi Leonov, to this day is something to behold.

But when that Cold War ended, we were rightly concerned that a lot of those weapons were going to get out into the wrong hands, a lot of that technology was going to get out into the wrong hands. And so to keep a lot of those former Soviet, now Russian, engineers working, there was this extraordinarily successful program of the Nunn-Lugar effort to go in to gather up those nuclear weapons and simultaneously to support the Russian aerospace industry and to buy this incredible engine, the RD–180. And so today, those engines play a significant role in meeting our Nation's launch requirements. We have already launched four missions this year alone using that engine.

And so now it is time that we are going to have to consider an alternative. Several of us in the Armed Services Committee put $100 million into the defense authorization bill to get that process started in this coming fiscal year. We want to make sure that the taxpayer money is well spent, and so it is important that we consider the launch needs with the goal in mind that we want assured access to space. And this is, obviously, not going to just affect the Department of Defense, although the National security activities are paramount. It clearly is going to involve commercial space activities as well and the question of preserving an industrial base.

And so the two committees represented at this dais have asked officials from DOD and NASA, along with many others, to come and discuss this issue of U.S. assured access to space. NASA has no stated need for a new engine and is already building its own space launch system. However, NASA obviously has extensive experience in building launch systems and is getting great experience in public/private partnerships. And so we are going to hear from all of these people.

Now, I am going to short circuit my remarks because we are racing the clock. We have a 10:15 vote and then a 12:20 vote. I am going to call on the chairman of the DOD defense authorization subcommittee and the ranking members to give some brief opening remarks, and then we will get into the panel. I am going to ask the panel to keep—your written comments are entered as a part of the record. I am going to ask you to keep it to about 3 minutes each so that we can then get into questions.

So, Senator Udall.

STATEMENT OF SENATOR MARK UDALL

Senator UDALL. Thank you, Senator Nelson.

In the spirit of Senator Nelson's comments, let me introduce my statement into the record and then make a couple comments on procedure. Given the number of witnesses and possible member attendance, I would propose to my colleagues that we use 5-minute rounds of questions.

And then as Senator Nelson pointed out, according to the floor staff, we have a vote at approximately 10:15 and another at 12:20. That being the case, I would like to ask that some of my colleagues remain to continue the hearing during the 10:15 vote while others vote and come back to switch places with them so that they may also go vote. And then we can repeat that procedure for the 12:20 vote if it is needed.

So again, I share Senator Nelson's sentiments. It is a very important hearing. I want to thank all of our witnesses for being here today.

Senator Nelson?

[The prepared statement of Senator Udall follows:]

[SUBCOMMITTEE INSERT]

Senator NELSON. Senator Sessions?

STATEMENT OF SENATOR JEFF SESSIONS

Senator SESSIONS. Thank you, Senator Nelson, for your observations and your opening statement.

Well, we are dealing with an important issue. It was not long ago that the Russian Deputy Prime Minister Rogozin stated this. After analyzing the sanctions against our space industry, I suggest the USA to bring their astronauts to the International Space Station using a trampoline. Close quote.

So we do not have assured access to space, as Senator Nelson has raised, and we have to have that. And I wish we were not in this situation. I wish we could have avoided it. So we are not and we need to make some changes.

The House has proposed legislation and identified $220 million in their authorization in appropriations committees to deal with the problem of developing a new rocket engine, which we can do. I am very confident about that. And I believe the price is going to be within our reach. Our committee has recommended $120 million. So we need to work on that. We need to see if NASA, Mr. Lightfoot, can contribute in this process.

So, Mr. Chairman, thank you for having the hearing. I think both committees are—it is good for us to be together. There are going to be some complexities, but I believe both houses of Congress have already laid out proposals that could work. And we have an excellent panel to help us make the right decision as we go forward. So thank you.

Senator NELSON. Thank you, Senator Sessions.

Senator Cruz.

Senator CRUZ. Thank you very much, Mr. Chairman.

I would like to begin by thanking the members of this panel for your service to this country and your efforts to ensure that the United States maintains a strong and capable space presence. The breadth of experience represented by this panel is impressive, and I appreciate your individual contributions towards America's national security.

I also want to thank members of the Armed Services Committee and the Commerce Committee for recognizing the need to hold a hearing on this issue and its impact on our country's access to space. It remains a simple reality that we need to work closely with the international community to guarantee that the International Space Station, its mission, and its crew are positively impacted by the decisions made here in Congress. Our astronauts and their peers are relying on a stable partnership to ensure their success.

The block purchase of 36 evolved expendable launch vehicle cores last year may have made economic sense during the global environment at that time and resulted in a meaningful savings, $4.4 billion, to the American taxpayer.

Although well intentioned, the unintended consequences of relying on a foreign supplier for critical national security equipment is now strikingly apparent. The United States is scrambling to maintain access to space and has no immediate options if the current supplier in Russia decides to cease export or if geopolitical circumstances dictate that the U.S. is no longer able to engage in a partnership with its supplier.

When the United States decided to utilize a foreign engine, RD–180, to boost our rockets into space, it was also agreed that production of that engine would ultimately occur in the United States. For whatever reason, whether it was for economic reasons or inattention, this never occurred. We find ourselves in this position as a result of our own inaction.

The United States must now respond decisively and provide the domestic capacity to launch both crew and cargo into space. The cost estimates for the design, construction, and testing and certification of a new multi-core engine are staggering in today's climate of limited financial resources. But we simply cannot rely on the vicissitudes of a foreign supplier in a foreign nation for our National security, and therefore we must do what it will take to reduce our reliance on foreign engines.

I look forward to hearing your suggestions, hearing your expertise as we work together on how best to alleviate this issue and defend the interest of the United States.

Thank you, Mr. Chairman.

Senator NELSON. Thank you, Senator Cruz.

Senators, rather than calling on you now, what I will do is I will forego my questions so we can get directly to you after we have heard from the witnesses.

So we have the Honorable Alan Estevez, Principal Deputy Under Secretary of Defense for Acquisition, Technology and Logistics. His testimony is going to focus on the current launch portfolio and the efforts to encourage competition and the options.

U.S. Air Force General William Shelton, Commander of Air Force Space Command. He will touch on the requirements for launching national security payloads, as well as the challenges presented with the RD–180.

Robert Lightfoot, NASA Associate Administrator. He will talk about NASA's launch requirements.

Ms. Cristina Chaplain, Director of Acquisition and Sourcing Management at the GAO. And she will discuss the efforts to encourage competition among the Government's launch services.

Retired U.S. Air Force Major General Howard Mitchell, Vice President for Program Assessments at the Aerospace Corporation.

Mr. Daniel Dumbacher, formerly NASA's Deputy Associate Administrator for Exploration Systems Development now at Purdue.

And Dr. Yool Kim, Senior Engineer at the RAND Corporation, will draw on assessment of risk related to the RD–180.

So I welcome all of you on behalf of the Senate, and we will start with you. I know it is compressed to get 3 minutes, but because of the interruption of votes today, it is of necessity and we want to get to questions. Mr. Estevez?

STATEMENT OF HON. ALAN F. ESTEVEZ, PRINCIPAL DEPUTY UNDER SECRETARY OF DEFENSE FOR ACQUISITION, TECHNOLOGY AND LOGISTICS

Mr. ESTEVEZ. Thank you, Senator Nelson.

Chairmen Nelson and Udall, Ranking Members Sessions and Cruz, distinguished members of the committee, I appreciate the opportunity to testify about assuring space access. I want to thank the committees for your providing support for our space-based capabilities. My written testimony has more detail, as you noted, and I ask that it be admitted to the record.

Defense space capabilities are central to our National security. Our assured access to space provides leaders and our men and women in uniform with unprecedented advantages in decision-making, military operations, and homeland security.

Since 2002, the Department of Defense has conducted 72 successful evolved expendable launch vehicle, or EELV, missions after refocusing on the importance of mission assurance following a string of failures in the 1990s.

To address concerns over the escalating costs of our National security space launch program, the Department restructured the EELV program in 2012. The restructured program balances efficient procurement of launch services, maintains the focus on mission assurance, and reintroduces competition into the EELV program. The restructured program also enabled the Air Force to award the contract for multiple launch services over a 5-year period. The contract helped stabilize the U.S. launch industrial base, saves the DOD and taxpayers more than $4.4 billion.

To facilitate competition going forward, the program is working with multiple potential new entrants launch service providers to successfully complete the new entrant certification process. The first new entrants could be certified later this year.

Years ago, we chose to utilize the Atlas V with the Russian RD–180 engine as a cost-effective way to meet space launch needs. However, the United States is not dependent on Russian technology to launch our critical space assets. The Delta IV launch vehicle has a domestically produced propulsion system that is capable of lifting all national security payloads. Once certified, new entrants are also expected to be able to lift a portion of the National security manifest using domestically produced propulsion systems. Today the Atlas V contractor, United Launch Alliance, maintains a Reserve stock of RD–180 engines in the United States and will support launches through late fiscal year 2016. Nevertheless, the long-term U.S. national security interests would be enhanced by shifting from the RD–180 to next generation U.S. engines in the most efficient and affordable manner.

The goal of the Department continues to be making space lift more affordable while reaching the advantages of competition. We have implemented the principles of better buying power, saving $4.4 billion, and have set in motion a sound strategy to foster future competition. In addition, the Department will continue to work with our interagency partners in creating an affordable, low-risk plan to reduce the Nation's reliance on Russian-manufactured propulsion systems.

Thank you for the opportunity to discuss our Nation's space launch capability. I look forward to your questions.

[The prepared statement of Mr. Estevez follows:]

Senator NELSON. Thank you.

General Shelton?

STATEMENT OF GEN. WILLIAM L. SHELTON, USAF, COMMANDER, AIR FORCE SPACE COMMAND

General SHELTON. Chairman Nelson, Chairman Udall, Senator Cruz, Senator Sessions, and distinguished members of both committees, it is a pleasure to represent Air Force Space Command here today.

It is also my privilege to appear with distinguished colleagues on the panel.

The Air Force's space capabilities are foundational to the joint warfighter and the Nation's capabilities who collectively rely on these systems across the range of civil and military operations. It is critical then that we ensure space services continue to be available at the times and places of our choosing, even in an increasingly challenged space domain. Ensuring these space services continue to be available starts with assured access to space.

Our ultimate objective is to safely and reliably place national security payloads on a schedule determined by the needs of the National security space enterprise. We are proud to have established an unprecedented launch success record with our evolved expendable launch vehicle program by placing an uncompromising premium on mission assurance.

Additionally, we have worked hard to reduce costs in our acquisition strategy with our current provider, ULA, and by progressively introducing competition into the launch business. But we must continue to insist on thorough, system engineering-based mission assurance processes. The loss of even one national security payload, both in terms of financial loss and operational impact, would make our mission insurance costs look like very cheap insurance. To make sure we sustain our incredible track record of success, we will continue to treat each and every launch as if it is our first.

Commensurate with the EELV's success, the commercial space launch industry has made substantial progress over the last year, including successful launches by Orbital Sciences and SpaceX. As a result, we are managing change in the EELV program from a single-provider environment to a multi-provider environment through a disciplined certification process. Through this process, we will continue to carefully and conservatively manage the introduction of full and open competition to ensure planned and future missions are delivered safely, successfully, and on schedule.

I thank you for your support, and I look forward to working with the Congress to provide resilient, capable, and affordable space capabilities for the joint force and for the Nation. Thank you.

[The prepared statement of General Shelton follows:]

Senator NELSON. Thank you.

Mr. Lightfoot?

STATEMENT OF ROBERT M. LIGHTFOOT, JR., ASSOCIATE ADMINISTRATOR, NATIONAL AERONAUTICS AND SPACE ADMINISTRATION

Mr. LIGHTFOOT. Thank you, Chairman Nelson, Chairman Udall, and other distinguished committee members. I appreciate the opportunity to testify before you on NASA's plans for ensuring access to space. My written testimony has been submitted for the record.

As you know, NASA has embarked on an ambitious path to send humans to Mars. This path includes conducting research aboard the International Space Station, developing the space launch system, Orion crew vehicle, and testing our new capabilities in the proving ground of cis-lunar space. We continue to do this with the cooperation from my international partner community.

As a critical element in this long-term exploration strategy, expanding commercial access to low earth orbit, or LEO as we call it, and extensive utilization of the International Space Station are among NASA's highest priorities. We will rely on and partner with U.S. industry and international partners for access to ISS while seeking to encourage innovation and to maintain a competitive environment for these services.

NASA continues to make strong progress on the space launch system, an exploration class heavy-lift launch vehicle designed for missions far beyond low earth orbit. SLS will begin with a lift capability of 70 metric tons, evolving to 105 metric tons, and eventually up to 130 metric tons. Near-term human exploration missions will benefit most from an enhanced upper stage. Increased booster thrust performance is not required until NASA undertakes more significant human missions such as landing on the surface of Mars. Our current needs do not require or have a need for a new LOX/hydrocarbon booster engine risk reduction or development effort at this time.

Through fiscal year 2020, NASA has plans to launch over 18 science missions of various size classes. We anticipate that our commercial launch service providers will add additional launch vehicles to our NASA launch services contract at some point in the near future.

NASA currently plans to launch payloads on six commercially provided Atlas V rockets which rely on the Russian-supplied RD–180 engines. Should the supply of these engines be disrupted, an interagency discussion would be required in order to allocate the available remaining RD–180s among national security and NASA considerations. Other launch vehicles would need to be considered using the appropriate procurement processes that we have in place.

NASA continues to work with our partners in the Department of Defense as it assesses approaches that could increase production rates and potentially reduce costs for launch systems that do not rely on the RD–180. We are committed to working with our partners to provider safe, reliable, and cost-effective access to space.

Mr. Chairman, thank you again for the opportunity to appear today, and I would be happy to respond to any questions you may have.

[The prepared statement of Mr. Lightfoot follows:]

Senator NELSON. Thank you.

Ms. Chaplain?

STATEMENT OF CRISTINA T. CHAPLAIN, DIRECTOR, ACQUISITION AND SOURCING MANAGEMENT, GOVERNMENT ACCOUNTABILITY OFFICE

Ms. CHAPLAIN. Chairman Udall, Chairman Nelson, Ranking Member Cruz, Ranking Member Sessions, and distinguished members of the committees, thank you for inviting me to participate in today's hearing.

I would just like to make three points about our relevant work.

First, as our testimony indicates, in the past we have highlighted deficiencies in the management and oversight of the EELV, as well as gaps in knowledge needed for the block buy that is now in place. Both DOD and the Congress have taken significant steps to rectify the problems we identified. For instance, DOD undertook rigorous efforts to obtain greater insight into program costs in advance of its contract negotiations with ULA. It took steps to reinstitute oversight reporting for the program, and it completed a new cost estimate. Over time, DOD has also come to recognize the value of competition for the EELV program, noting that with no threat of competition DOD was in a poor negotiating position.

Second, with respect to the current competition, we have reported on the benefits and the challenges associated with how DOD could run the competition, but we did not recommend a specific approach as that decision should be made by DOD based on its requirements and resources. Important factors include the need to maintain a high degree of reliability, as the satellites being launched are expensive and vital to national security, the need for flexibility in launch scheduling, the importance of retaining cost and pricing data, the need to keep costs down, and considerations about the Government's future demand for launch services.

Third, my testimony identifies best practices that should be adopted in future rocket engine or launch vehicle development efforts. The one I would like to stress here is the need for decisions to be made with a Government-wide perspective and a long-term perspective. Our work has shown that defense and civilian Government agencies together expect to require nearly $44 billion for the next 5 years for launch activities. At the same time, our past work has found that launch acquisitions and activities have not always been well coordinated, though DOD and NASA have made progress on that front since then. Concerns have also been raised in various studies about the lack of strategic planning and investment for future technologies. Further, industry is at a crossroad with new vendors emerging and certain strategic capabilities and less demand by the Government. The bottom line is that any new launch vehicle effort is likely to have impacts that reach beyond DOD and the

EELV program and should be carefully considered in a Government-wide and long-term context.

This concludes my statement. I look forward to answering your questions.

[The prepared statement of Ms. Chaplain follows:]

Senator NELSON. Thank you.

General Mitchell?

STATEMENT OF MAJ. GEN. HOWARD J. MITCHELL, USAF (RET.), VICE PRESIDENT, PROGRAM ASSESSMENTS, THE AEROSPACE CORPORATION

Mr. MITCHELL. Chairman Nelson, Chairman Udall, thank you for this opportunity to speak and I also thank the rest of the members that are attending the hearing.

I was asked to do a study on the RD–180 mitigation study. I recently chaired that under a terms of reference that was signed by the Assistant Secretary of the Air Force for Acquisition. I have provided the committee a version of that briefing for the record, as well as my opening comments.

I will just hit the four major areas that the study identified.

First, a disruption of the RD–180 engine could have significant impact on the United States' ability to launch DOD, intelligence community, NASA, NOAA, and commercial satellites scheduled to launch on Atlas V through 2020. Neither the Delta IV nor new entrants can help mitigate that impact until 2017 and beyond.

Second, there are several upcoming events that bear monitoring as they can provide indications of the Russians' intent and the United States' intent.

Third, the current Air Force strategy for competition can be adversely affected should the Atlas not be available for competition.

And fourth, in the 2022–2023 time frame with appropriate near-term funding for technology maturation, the Nation could have new launch capabilities based on liquid oxygen/hydrogen engine technology that do not rely on foreign sources.

I look forward to answering any questions you might have.

[The prepared statement of Mr. Mitchell follows:]

Senator NELSON. Thank you, General.

Mr. Dumbacher?

STATEMENT OF DANIEL L. DUMBACHER, PROFESSOR OF PRACTICE, DEPARTMENT OF AERONAUTICS AND AERO-SPACE ENGINEERING, PURDUE UNIVERSITY

Mr. DUMBACHER. Chairman Nelson, Chairman Udall, and members of today's respective committees, thank you for the opportunity to discuss the current state of the U.S. launch enterprise.

The United States' ability to achieve and go beyond low earth orbit is essential for our Nation's defense, commercial, and space exploration enterprises.

Leaving the surface of the earth and attaining orbital velocity at 17,500 miles per hour is a complex systems challenge. Factors key to achieving this task are mission requirements, technical performance, development risk and cost, operations cost, schedule, industrial base, and yes, even political concerns, which all must be addressed with multiple stakeholders.

In the early phase of a rocket launch, thrust is more important to initially overcome the Earth's gravity than propulsion efficiency. However, as the vehicle progresses to higher altitudes and climbs out of the Earth's gravity well, propulsion efficiency becomes more important, even as thrust remains an important technical parameter.

When NASA was preparing to go to the moon in the 1960s, it determined that large amounts of thrust were needed for the first 2½ minutes of flight to put the Apollo spacecraft and lunar lander on the surface of the moon. To meet the mission need, NASA recognized that much development and testing effort of liquid oxygen/kerosene systems was required and therefore restarted the Air Force's E–1 development from the 1950s as the F–1 program.

During the development of the Space Shuttle, NASA determined that it had a lower payload delivery requirement and constrained budgets. The development cost estimates for the shuttle's solid booster were lower than competing booster liquid systems. Many of the same challenges were again considered by NASA during the planning and development process for the space launch system. NASA assessed many launch configurations, weighing the pros and cons of each. Again, technical performance, challenges associated with limited budgets, the need to launch the first flight as early as possible, and impacts to the propulsion industrial base weighed heavily on NASA's decision-making.

Ultimately for the space launch system, NASA determined that using the solid boosters based on shuttle and constellation experience minimized the upfront development costs, reduced the development risks, and most likely would result in a more timely first flight of the space launch system. NASA also chose to utilize over 40 years of investment in large liquid oxygen/liquid hydrogen engines to minimize development cost and risk.

Following the Apollo program, the U.S. Government dramatically limited its hydrocarbon investments and focused on utilizing solid propulsion systems and liquid oxygen/liquid hydrogen systems. The United States leads the world in these propulsion systems. However, we need to reduce the costs of these systems. In my opinion, the United States should build upon its long investment in solid and liquid oxygen/liquid hydrogen propulsion systems and allow the marketplace to provide viable choices for use by NASA and the Department of Defense. Competition will incentivize industry to develop efficient management models, use the new technologies that will reduce costs, and continue to search for and develop technologies necessary to reduce development and operations costs.

Thank you for allowing me to appear before you today. More details are included in the submitted written testimony. I will be happy to answer your questions.

[The prepared statement of Mr. Dumbacher follows:]

Senator NELSON. Thank you.

Dr. Kim?

STATEMENT OF DR. YOOL KIM, SENIOR ENGINEER, THE RAND CORPORATION

Dr. YIM. Mr. Chairmen, ranking members, and distinguished committee and subcommittee members, thank you for the opportunity to testify before you today on this important issue.

Today I will focus on the conclusions of a RAND study mandated by Congress on the National security implications of continuing to use foreign components for launch vehicles under the evolved expendable launch vehicle program. I will discuss risks of using foreign components under the EELV program and the potential effects on the U.S. space launch capability and national security space missions if an interruption in the supply of those components should occur.

The Atlas V and the Delta IV launch vehicles in the EELV program have several major foreign components or subsystems and many more lower tier components from countries all over the world. The risk of potential supply interruption of most foreign components in the EELV program is low and manageable.

The foreign component of most concern is the Russian RD–180 engine, the primary booster of the Atlas V launch vehicle. The RD–180 engine supplier poses a moderate risk of a supply interruption primarily related to the political concerns with Russia, although the supplier has strong financial incentives to continue deliveries to United Launch Alliance. The RD–180 engine is the most critical component in terms of costs, schedule, and the technical difficulty associated with developing an alternative engine source. An interruption in the RD–180 engine supply would cause a significant disruption in EELV launch operations because a large number of Atlas V launches are scheduled in the next few years, and a significant effort would have to be made to mitigate the disruption.

Should a long-term interruption in the RD–180 engine supply occur, risks to the U.S. space launch capability could be mitigated by using the stockpile of RD–180 engines that United Launch Alliance maintains and by moving some satellites carried on the Atlas V to Delta IV until a new entrant launch vehicle from a different launch service provider or a re-engined Atlas V becomes available. However, the mitigation efforts have significant costs implications and relying on a single launch vehicle would pose a higher risk to U.S. access to space. More details on the mitigation approach and remaining risks are described in my written testimony.

Although some national security space satellites are likely to be delayed during disruption, the risk will be low for most national security space missions if national security space satellites are given priority in use of the RD–180 engine stockpile, particularly for the launch of the critical satellites on Atlas V. However, many variables will influence the mitigation approach which should be based on a consideration of the tradeoffs regarding the cost and schedule and the mission risks of different options.

Again, thank you for inviting me here today to testify on this very important national issue. I look forward to your questions.

[The prepared statement of Dr. Yim follows:]

Senator NELSON. Thank you.

Congratulations, you have set a Senate record. [Laughter.]

All seven of you. And so it leaves time for our Senators to ask. I will defer my questions and do cleanup at the end. And we will call—other than the chair and the ranking, we will call the Senators in order in which they came. So, Senator Udall?

Senator UDALL. Thank you, Mr. Chairman.

Our mission here as a Congress and a country is to assure access to space for both civil and military missions. And both committees are here because this is so vital. Space access is vital to our economy, our National security, as well as communications, our weather forecast networks, and scientific efforts. And in that context, I want to just touch on two of, I think, key issues that are attached to this mission.

First is the ongoing effort to introduce competition into the launch market. Having additional certified competitors in the marketplace will help lower the cost of delivering payloads into space and, of course, drive innovation. We must also ensure that those providers, though—and many of you have spoken to this—are able to meet the technical requirements necessary to provide mission assurance. We make significant investments as a country in our space-based assets and we have got to be absolutely confident that they reach the proper orbit safely.

And second, we have got to address the recent developments with Russia and our reliance on the Russian-built liquid rocket engine used on the Atlas V medium-lift vehicle. Atlas, as you all know, is a proven workhorse with a tremendous record of success for civilian and military lift. Since the 1990s, our policy has been to stockpile the Russian engine rather than develop a domestic engine. We are now reevaluating that policy, and you all have begun to give us a sense of how we should proceed.

In that context, I want to turn to General Shelton, who is known as someone who will give us frank, no-nonsense assessments, and we truly appreciate your contributions in leadership, General Shelton. I know you are going to retire soon. I am particularly glad you are going to stay in Colorado Springs, and I look forward to working with you whatever your retirement holds because I know you and I know you are not really going to retire. So let me turn to you.

You place, as the Air Force, high value on mission assurance, given the critical importance of cost of the satellites that DOD launches. I have got a multi-part question related to mission assurance.

Number one, can you explain what effects the launch failures in the 1990s had on the DOD? That is number one.

Number two, can you explain the role of mission assurance in the current EELV program as compared to the late 1990s when those failures occurred?

And can you, finally, explain the importance of the added performance margin the Air Force puts on the EELV rockets and how that margin contributes to mission assurance?

General SHELTON. Thank you, Senator Udall. Thank you for your kind words too, of course.

Senator UDALL. Well deserved.

General SHELTON. As we look back at the 1990s, between 1997 and 1999, we had significant failures both on the military side and the commercial side, including three Titan IVs which launched our most significant payloads at that time.

We had adjusted our approach to mission assurance from what has been traditionally deep oversight into just insight. So we pretty much gave it over to the contractors to provide their own mission assurance. And we found out that just did not work well for us.

So as we turned that around and went through some extensive introspection, had a nationally significant study come forward, we decided to get back into the deep oversight business, and that is what we do today. Very deep penetration of process, very deep penetration of actual processing of every launch vehicle. And as I said in my opening statement, we treat every launch as if it is our very first, and so what happened in the past in terms of success, we do not pay much attention to. We pay a lot of attention to the launch of the day.

As we look at that performance margin, 7 percent, about 5 percent of that—or 5 of the 7 is for mission growth. We order launch vehicles about 2 years in advance. So we see payloads sometimes get heavier during their development process toward the end, and we Reserve that 5 percent. The additional 2 percent is just in case something goes wrong with that rocket. So about a year and a half ago, we were launching a GPS mission, and we had problems with the upper stage. Luckily we had margin to make it to orbit or we would have had a failed mission. So that is the reason for that margin. That is the reason for our continued emphasis on mission assurance.

Senator UDALL. In your opinion, do we need to develop a liquid rocket engine for medium and heavy lift, and if so, how urgent is that requirement?

General SHELTON. If you look at what has happened to us now in the last few months, I think it points to a vulnerability that we have. We had decided to rely on a foreign supplier. Probably the most advanced rocket engine in the world, by the way. And that has worked extremely well. If you look at the Atlas V performance, there is nothing to complain about the Atlas V performance.

But given that reliance, it is probably time to look at strategies for the future, and I think we can certainly help our liquid rocket engine industrial base by moving into such a program. I think we need to study the requirements. I think we need to look at what kinds of technologies we need to develop, but in my opinion, it is time to move off reliance on that foreign engine.

Senator UDALL. Thank you.

Senator NELSON. Senator Sessions?

Senator SESSIONS. Thank you. It is good to follow my chairman, Senator Udall.

And to follow up on that, General Shelton, in your opinion is it a national security priority for the United States to develop an American-made engine that could replace the RD–180 first?

General SHELTON. Yes, sir. If you consider space a national security priority, then you absolutely have to consider assured access to space a national security priority. So given that we have a vulnerability here, it is time to close that hole.

Senator SESSIONS. I could not agree more. We definitely depend on space capability for communications, for observation, and it is just a base part of our National security, as well as our commercial activity.

Mr. Rogozin, a deputy prime minister in Russia, also said in May, quote, Russia is ready to continue deliveries of RD–180 engines to the United States only under the guarantee that they will not be used in the interest of the Pentagon. Close quote. You are part of the Pentagon, are you not?

General SHELTON. Yes, sir.

Senator SESSIONS. Well, look, that is just not acceptable, and it puts us in a vulnerable position that I wish we did not have to be in, but it is time for us to rise to the occasion and fix this situation. I am open to ways to do it, and we will keep working to do it in a way that is effective.

But I have been pleased with the Senate legislation that we worked on. I think that balanced and considered the challenges that we faced and tried to do it in the right way. The House has also come up with a similar proposal. So I am glad to have a public hearing about this and discuss it. Let us just keep talking about it to try to get it right.

Mr. Lightfoot, you have been at NASA for some time, and you started the National Institute for Rocket Propulsion Systems and have studied these issues over the years. Does the institute at NASA have the people with the skills and experience that could assist the Air Force in this effort to develop an American replacement for this engine? I guess I would ask, if so, would NASA desire to be compensated for their efforts?

Mr. LIGHTFOOT. Well, Senator Sessions, I think the National Institute of Rocket Propulsion Systems that we put in place is not just NASA. The DOD is part of that, as well as other agencies and a lot of the industry folks. So I think the institute itself was set up to be able to pull together all the propulsion system resources this country has to solve problems that could come up, whatever they are. So if we chose to go down that path, I surely would think we would use—NIRPS, as we call it, it would be part of that solution space. And you get to pull in the expertise that all the Government agencies have if we choose to go that way.

That is not in our plan today. So, of course, we would be interested in getting compensated for that. But I think the team right now works together fairly well with us and the DOD from that standpoint.

Senator SESSIONS. And you do have something, you believe, at NASA, and NASA has something they could contribute to the effort.

Mr. LIGHTFOOT. Yes. I think our expertise that has been hands-on for years in developing our own launch systems we can bring to that story as well.

The issue that Mr. Dumbacher talked about in terms of the lack of development of LOX/hydrocarbon engines in the past—this is a gap in our base, but the team at Marshall Space Flight Center in particular has been working on this for some time in a low level activity to try to keep up with that technology as we move forward. And I think we can bring that to bear to help our friends out if we choose to go that way.

Senator SESSIONS. Mr. Dumbacher, I will ask you. You are more of an independent observer here perhaps. But do you think that there will be any fundamental technological engineering difficulties that would make it hard or unlikely that we could develop this engine? Or do you believe the United States could produce an engine similar to the RD–180 that could be as effective or more so?

Mr. DUMBACHER. Senator, I believe that we can do that, that we can develop an engine within a sufficient time and money. There are development risks associated with it, three in particular. One is the high pressure oxygen compatibility of the materials that we use in the engine system. Combustion stability has been an historic issue with large LOX/kerosene engines, and also how you handle the start transient depending upon which cycle you use for those engines. So there are technical issues to be addressed. I think we can overcome those, but it is a matter of time and money required to do that.

Senator SESSIONS. Mr. Lightfoot, do you think it is also technologically—it seems to me it is a fairly mature technology now. How do you feel about it? What kind of confidence level do you have that such an engine could be produced?

Mr. LIGHTFOOT. Well, sir, I think that we could get to an engine, as Dan said. I think the challenge we have to look at is the launch system is not just the engine. There will be impacts that go to the launch vehicle, to the launch infrastructure. When we talk about a launch capability, I have an analogy I have been using related to—I have hybrid cars, 4-cylinder, 6-cylinder, 8-cylinder, and I have diesels. If I change your engine, I would probably change your car a little bit as well. So we have to look at the impacts on not just building an engine and having an engine to use, but the impacts on the infrastructure that goes around that. Can we do it? I am sure we can with the right resources to go do it. But I think we have to make sure we understand the other pieces that come with that as we go forward.

Senator SESSIONS. My time is up.

Senator NELSON. Senator Cruz?

Senator CRUZ. Thank you, Mr. Chairman. Thank you, all of you, for coming here and providing your expertise.

In life, it is often a wise strategy to hope for the best and yet plan for the worst. And that is especially true in the National security context. So I would welcome the wisdom of this panel on what the implications would be for our National security if the worst occurs and what the best avenues we have to alleviate those implications.

And in particular, General Shelton, I would like to start with you. Assume that conditions and relations with Russia deteriorate substantially. I hope that they do not, but assume that they do. Assume that Putin adopts a position of maximum belligerence and picks up the phone and instructs all engine exports end tomorrow. What would the implications be for U.S. national security if that decision were made?

General SHELTON. Senator, as we look at that, as one of the number of scenarios we have considered, we think as a minimum that would be about $1.5 billion. We think that that would stretch out launches. We have got to ramp up the production of our Delta factory, which would take some time. So that would stretch out launches maybe 12 to 20 months in some cases; for the heavier missions, maybe even 48 months. So that puts constellations at risk, and the ones that we are talking about, the heaviest ones, are our most significant constellations. So it is dire. If that should happen, there is no question that inside this manifest that we are considering right now, there would be serious national security implications.

Senator CRUZ. And to what extent could our existing stockpile of engines reasonably be stretched to cover the needs? How long could we expect it to cover our needs?

General SHELTON. Sir, we have 15 engines left right now. Depending on how we chose to meter those out, I think Mr. Lightfoot said in his opening statement that we would have to meet nationally to decide how we would allocate those engines. But that is all part of this set of scenarios that we are considering right now. We do not know the exact impact until we get together and decide how we would allocate those 15.

Senator CRUZ. On best case, are we talking a year? Are we talking 2 years? How long would you reasonably expect the stockpile to be able to meet the needs?

General SHELTON. We could meter that out over a number of years depending on what you decided to spend those engines on.

Senator CRUZ. Mr. Lightfoot, let me ask you the same hypothetical, maximum belligerence, but let us assume that Mr. Putin did not just say engine exports. He also said the Soyuz has a shutdown. No more Americans will have access to our launch capacity for manned launch. What would the impact be on the Space Station and on our needs?

Mr. LIGHTFOOT. So I think we would clearly have to go assess what that would do to us from access to station. To date, we have seen no change in the behaviors at all. We continue to launch Soyuz and put people up there. I want to make sure that is really clear regardless of Rogozin's comments. Our teams are working together with the Russians very well to continue the Space Station operations. Clearly the International Space Station is our stepping stone to our larger exploration program. So we would have to go look at the implications associated with that, and it would be significant from that standpoint. And then we would work—

Senator CRUZ. Can you briefly describe, if it was cut off, what the implications would be in your best judgment?

Mr. LIGHTFOOT. Yes. We would want to accelerate our commercial crew activity so that we are launching from here, from the U.S. from that standpoint, and let our partners that have bid on that proposal now and the one that we have in selection—let them come forward and show us how they would provide the access to the International Space Station that we need.

But it is more than just the launch to the station. The Russians also operate key components on the station just like we do for them. So that would be the issues that we would have to go assess kind of one by one.

Senator CRUZ. Well, in light of those concerns, both with engines and manned launch capacity, if a decision were made to proceed forward with maximum speed towards acquiring the domestic capacity to fill these needs, realistically how quickly could we do so and how many commercial enterprises possess the skill and expertise to be credibly able to meet that need?

General SHELTON. I will start.

The only provider that is really in a serious certification process right now is SpaceX. If everything goes extremely well, a very green light schedule, by December of this year, we could have them certified. If you look at Atlas V, there are 10 configurations of Atlas V due to the various upper stages, strap-on solids, those sorts of things. SpaceX—there are seven of those 10 configurations that they could not launch. They do not have the lift capacity for that. They have a heavy vehicle planned in the future, but that is down the road a ways. So that means that SpaceX could compete for some of those. We would need to ramp up Delta IV production to accommodate the rest. So we are probably looking at, again, 12 to 48 months slip on some launches.

Senator CRUZ. Mr. Lightfoot, anything to add on the manned launch?

Mr. LIGHTFOOT. No. I think we would work with these guys to figure out which critical missions we needed to get done and how we would work them into the manifest. But we should expect a significant impact.

Senator CRUZ. How quickly could we fill the manifest?

Mr. LIGHTFOOT. At the current proposals that we have, we think we can be flying by 2017, putting humans into space from our location. I do not know if there is much we can accelerate at this point because of the way the process works, but that is what we would prefer to do, is just keep the funding for the commercial crew program going so we can meet the 2017 date.

Senator CRUZ. Thank you very much.

Senator NELSON. Be prepared to answer the question, if you got more money, could you accelerate that to 2016.

Senator Wicker?

Senator WICKER. Well, why don?t you just answer that question? [Laughter.]

Mr. LIGHTFOOT. I think part of the issue we are dealing with is is we are in the middle of a procurement. Right? So we have a procurement right now that we will make a selection on later this fiscal year. And having not seen the proposals because I am not part of that procurement board, I cannot tell you what the acceleration options are. However, at some point, when you order—we are in 2014 already here. So when you order a rocket, we typically order them 3 years in advance from when we are doing it. So that is where we are.

Senator WICKER. So there are considerations other than funding that are going to take time.

Mr. LIGHTFOOT. Yes, sir. Yes, sir, exactly, to manufacture the long-lead items and all the different pieces that come into building these things.

Senator WICKER. Well, I think probably the chairman would want you to get back to us about how we can be helpful in pushing the timeline.

Let me ask this. I will start with General Shelton and Mr. Estevez. Now, in developing the U.S. alternative to the Russian

RD–180 engine, the Air Force research laboratories are going to want to be involved. And Stennis Space Center is going to be a key player. Can you tell us at this point how costs for testing compare between these two facilities? And given their respective workloads and priorities, would the developmental timeline and costs be less by utilizing the Stennis Space Center as opposed to the Air Force Research Laboratory?

General SHELTON. Sir, I could not give you an outright comparison, and I would not necessarily expect it would be an either/or. I would think that that would be divvied up. We have had work going on in the Air Force Research Laboratory facility since 2007 on the hydrocarbon engine. That was part of the direction coming out after the decision to not co-produce RD–180s and to stockpile instead. So I think Stennis has great capability. I think we would utilize Stennis for some things. I think we would use AFRL facilities for other things. And we would use commercial facilities for yet other things.

Senator WICKER. Okay. And before we move to Mr. Estevez, you could go back, though, and get the committee some cost comparisons for testing in the past and supply them to us on the record. Would you be able to do that?

General SHELTON. Yes, sir.

Senator WICKER. Look for that and let us know.

General SHELTON. We will try to do something that is apples to apples. That may be difficult but we will give it a shot.

[The information referred to follows:]

[SUBCOMMITTEE INSERT]

Senator WICKER. Okay.

Mr. Estevez?

Mr. ESTEVEZ. First, let me—I would reiterate what General Shelton said. But we are looking at this, you know, how to move forward with a replacement for the RD–180, as a whole-of-government issue. To Mr. Lightfoot's earlier comments, we would look at NASA capabilities, as well as what we have inside the Department of Defense and, as General Shelton said, look at what the commercial sector is also doing. So we have not decided what the best way forward is; ergo, it is preliminary to decide where we would start that development. We do want to move forward with some risk reduction activities. In fact, we put some money in the reprogramming action that we just put before Congress to do that.

Senator WICKER. Mr. Estevez, do you agree, though, with the General that Stennis and the Research Laboratory would be key players——

Mr. ESTEVEZ. I do.

Senator WICKER.—in any path forward?

Mr. ESTEVEZ. I do.

Senator WICKER. Okay. Now, for all of you, who wants to volunteer here? GAO has argued that there is room for improvement in coordination between NASA and DOD for future programs. Who has seen this report, and what do you believe could be done to make the improvements that GAO has suggested would need to be made? Who wants to tackle that? Mr. Lightfoot?

Mr. LIGHTFOOT. I will start. Then I will let these guys jump in. We have done several things to——

Senator WICKER. Do they have a point?

Mr. LIGHTFOOT. I think when the report came out, they had a point. I think we have done a lot since then, though, to improve that communication.

Senator WICKER. Already?

Mr. LIGHTFOOT. Yes, sir. We both have folks on site at all of our new entrants and the ULA folks as well, and we share the information, as best we can, across with each other depending on where the certification process is between the launch vehicles. So we are going through that process and sharing information, which we think is the most important thing as we move through certification of these new entrants in the process.

General SHELTON. Senator, I believe there is tremendous transparency between NASA and the Air Force. There are processes set up to ensure that we are communicating back and forth. There are summits that occur at NASA, Air Force, NRO leadership levels. It is hard to imagine it could be any better. We have different requirements.

Senator WICKER. So the GAO conclusion was perhaps a little unfair.

General SHELTON. Senator, that is not what I am saying. Things in the past—I would agree there were some areas we could do better in and we have.

Senator WICKER. Okay.

Mr. Estevez, let me ask you this. We are all agreed that there is a great opportunity for public/private partnerships in this engine development idea. Is that correct?

Mr. ESTEVEZ. Yes, sir. In fact, we are doing an assessment now about what course to take in a replacement for the RD–180. You can look at an inside-Government-only development. That is probably not the best course, but we are going to look at that, look at jump starting some things, you know, some risk reduction, and then turning that to the private sector, hopefully, that they will build. That is one way to do it. A public/private partnership is another way. And so we have an assessment, after the Mitchell study laid out some ways forward, that is ongoing inside the Air Force right now, and we expect that to come out sometime this fall.

Senator WICKER. General Mitchell, do you want to conclude my brief few seconds on that issue?

Mr. MITCHELL. Yes, sir. There are some risk reduction activities that need to bring the technology levels up in hydrocarbon engines that need to be invested in. That will take a year and a half or so or 2 years to bring those technologies up that were discussed. They have to do with the materials. They have to do with the modeling of the combustion instability and some of the piece parts of the engine itself, injectors and other components, that need to be matured over the next year-year and a half to position yourself to start a full-scale development program probably in the last fiscal year 2016-fiscal year 2017 time frame.

Senator WICKER. Thank you, Mr. Chairman.

Senator NELSON. I am going to recess the committee momentarily to go over and finish the vote. Chairman Udall will take over the committee when he arrives.

Let me just ask. I have seen the Air Force study on a replacement for the RD–180. I have heard various estimates on costs. I have heard various estimates on time. Are we looking at, in reality, 7 or 8 years to have an engine ready to go in a rocket, whether it be a version of the Atlas V or whatever? Is that a realistic time frame, or is it more or can it be less?

Mr. ESTEVEZ. Too soon to tell. 8 years is—you know, 5 to 8 years is kind of what we are looking at. We want to do this right, though, to the earlier discussion about mission assurance, and figuring out the course and what the most affordable way to do it is also a key part of that decision point. While this is a priority, there are other things in the mix. So I would hesitate to make a firm projection at this point, Senator, until we know exactly where we are going. We do know we are going to replace it, though.

Senator NELSON. Do you want to give a ball park on cost to replace the RD–180?

Mr. ESTEVEZ. Estimates were in the $1 billion to $2 billion. Again, until we get a course ahead how we are going to do that public/private partnership, Government-only, you know, all those things change that dynamic. So I would really hesitate to make a true assessment of that at this point.

Senator NELSON. And that is overall cost, including the alterations to the rocket that you would put it in.

Mr. ESTEVEZ. I would have to go back and take that for the record, Senator, but I am happy to give you an estimate there.

[The information referred to follows:]

[SUBCOMMITTEE INSERT]

Senator NELSON. Okay.

Well, the good news is that it seems like that President Putin is not quite as aggressive as he first appeared to be. It is also fairly clear that Roscosmos certainly does not want to give up that income stream, and it looks like that from their standpoint, they clearly want to continue to supply the RD–180.

But, you know, we have kind of seen this movie before. It was back a decade or so ago that we said we were going to start the process of the replacement of the RD–180, right when we started acquiring these in the 1990s, and then we backed off of that. So here we are again, and that is part of the issue of the day.

I am going to recess the committee because we are down to 1 minute to vote. I am going to see if I can sprint. The committee will stand in recess until the call of the chair. [Recess.}

Senator UDALL [presiding]. The committee will come to order. Thank you for your patience.

I want to recognize Senator Kaine for 5 minutes.

Senator KAINE. Thank you, Mr. Chairman, and thank you to the witnesses for your service and your testimony today.

I have questions that are primarily going to be addressed to General Shelton and Assistant Administrator Lightfoot on the Wallops facility in Virginia.

As you know, on—I guess it was last Sunday, July 13th, Orbital successfully launched the second resupply mission to the International Space Station from Wallops in Virginia. Orbital is lined up to carry out eight cargo space missions to the ISS through 2016, making it a critical player in commercial space. And Wallop is also in a unique position, and it is capable of launching certain national security payloads from that facility.

Administrator Lightfoot, what are some of the benefits of having facilities like Wallops for launching smaller and mid-size payloads?

Mr. LIGHTFOOT. Well, sir, I think you saw it today. The Cygnus spacecraft birthed to the International Space Station about 6:30 this morning. So it was pretty exciting for us to get the crews there.

The other thing it gives us is it gives us two different access points to get to the station, not only two different providers in SpaceX and Orbital, but we now have two locations from which to fly, which helps us from that standpoint. So it has been very good for us an agency to have the folks at Wallops and their little team do what they have done to provide us that access to the station.

Senator KAINE. Just on that, it is good to have two places to launch from. Is that just a matter just of scheduling, gives you more flexibility or also are there sort of aerospace reasons why launching sites in different parts of the country are helpful?

Mr. LIGHTFOOT. What it does is it gives the commercial providers—because we do not pick their launch site for them. It gives them the opportunity to get the best value for their launch vehicle, and so that is the advantage that they have had. We do not set where they fly from. They just go make that agreement themselves, and that makes it more of a competitive process.

Senator KAINE. What are we doing to ensure that national defense agencies have some redundant capability for launching national security payloads into space? In the event that problems at either Vandenberg or Canaveral would occur such as natural disasters, et cetera, I would imagine that redundancy is a positive. General Shelton?

General SHELTON. Senator, there are physics-based reasons for having two different launch locations. For example, going out of Vandenberg, you are looking at largely polar orbits or test activity that goes out to the west. From the Cape, you are looking at lower inclination orbits. And there really is not a way to produce redundancy for that physics-based problem unless you build brand new launch facilities. So, yes, we would be susceptible to a very broad destructive kind of event, but that has not happened in the history of space flight.

Senator KAINE. Let me move on to the RD–180 replacement. And, General Shelton, you may have answered questions on this already. Forgive me for coming in a few minutes late. But what can the DOD do to accelerate a timeline to develop a U.S.-built alternative to the Russian engine?

General SHELTON. Certainly it will take a very serious funding commitment, and we will go through some risk reduction efforts here and technology maturation efforts over the next couple of years. And then beyond that, it will take some very serious investment. If we want to stretch out the program, we can, but if we want to really get after a serious program, then we are going to have to have significant investment.

Senator KAINE. Thank you.

No other questions, Mr. Chairman. I just want to indicate that Wallops, the launch site in Virginia, is kind of a well-kept secret.

Most Virginians do not even know that rockets launch out of Wallops, which is just off the eastern shore, but more rockets have launched from Wallops than either Vandenberg or Cape Canaveral. The reason they are not generally known is that they are unmanned and not manned and they tend to be smaller, but it serves as a significant asset. We worked hard on it with our colleagues in Maryland because it is very close to the Maryland border. Chairwoman Mikulski has been a huge supporter of investments there. And that additional launch capacity I think has served the Nation well. And I look forward to working with you in the future to continue that.

Thank you, Mr. Chairman.

Senator UDALL. Thank you, Senator Kaine. And the Commonwealth never ceases to surprise all of us. [Laughter.]

I learned something as well. That is an important part of our whole aerospace consortium, if you will. Thank you for that.

Mr. Estevez, let me turn to you, if I might. Before the current 36-core block buy, we procured our space launch as a service using a commercial waiver under the Federal acquisition regulations that provided no cost insight into the structure of the procurement. What was the result of this waiver on that particular decision?

Mr. ESTEVEZ. The statute requires us to acquire space launch services under FAR, part 12, which is a commercial service. And there are good reasons for that, especially as we moved into commercial areas. However, we bought the block buy under FAR 15. So we have full cost and pricing data from ULA. It gives us great insight into the cost structure of that. Going forward, I am not sure——

Senator UDALL. Would you recommend having the EELV program use it again?

Mr. ESTEVEZ. Again, we are looking at the benefits and the negatives on that. Certain commercial providers do not have to deal with the same business system background and the like that we require under FAR, part 15. So it has to be weighed acquisition by acquisition frankly. But there are benefits to having that full cost and pricing that have been helpful to us.

Senator UDALL. Let me follow up with asking you to give us an explanation. Can you explain the nature of the cost overruns of that prior EELV contract? And then what cost savings were achieved in the current contract, and how were they obtained?

Mr. ESTEVEZ. The past was not really cost overruns because we were buying launch services as a contract by contract, launch by launch. So we are buying a one each. But if you look at the program overall for the depth of the manifest, there was great cost growth; ergo, we ran into a Nunn-McCurdy situation for the EELV program.

What we were able to achieve by doing the 36-core buy is economies of scale. So ULA could go to their industry subs and give them a deal because they know they are going to launch a certain number; ergo, it lowers the cost in total for that. It gave us price stability. It gave them an understanding of what their business base was going to be, $4.4 billion over our projections in fiscal year 2012 savings to the Department of Defense, to the American taxpayer. So a great benefit to us.

Senator UDALL. Let me turn to Ms. Chaplain. Nice to see you as always. I believe this is the third time you have testified this year before our subcommittee, which I believe qualifies you for frequent flyer status, whether on an airplane or rocket. You can maybe make the choice.

Can you explain why the waiver under the Federal acquisition regulations led to the lack of transparency in the cost increases in the EELV program?

Ms. CHAPLAIN. Yes. It is pretty simple. With the waiver, the Government did not have the type of underlying cost and pricing data on critical pieces like the engines that it needed to make good negotiations, especially as it was going to commit to a large span of time under the block buy. So without that kind of data and if you are in a sole source environment, you are really crippled in terms of your negotiating position. If there is a competitive environment, it might not be such an issue because the competition itself can drive down prices.

Senator UDALL. You have reviewed the 36-core block buy, the current core block buy. Do you agree with that estimated cost savings of $4 billion on the current EELV contract?

Ms. CHAPLAIN. So we have not thoroughly assessed the savings claim, but we do know that the Air Force took all the actions it needed to obtain those kinds of significant savings. So they did gain much more insight into cost and pricing. They went through their launch processes, understood them more. They understood pieces of cost better and were just able to account for more things. When they went to the bargaining table, they were in a much better bargaining position.

Senator UDALL. General Mitchell, let me turn to you, if I might. Thank you for your service both on active duty but also on the committee that you helmed.

You recommended the development of a domestic engine, I believe, to replace the RD–180. And I assume the committee reviewed proposals from industry. How mature are those proposals, and what are the major technical hurdles in their development?

Mr. MITCHELL. So I think we talked to all of the folks who had engine developments, and they range from what I will call viewgraphs to some piece parts that have been done to concepts. Nobody has all of the technology ready to start a full-scale development program in our review. So we think that that is going to take some investment and time to get the technologies up to where you could actually do what we call a full-scale development and commit to actually procuring the new engine.

Senator UDALL. So it is more than possible, but there is a significant amount of time between here and there.

Mr. MITCHELL. Yes, sir. And the areas primarily revolve around engine components, injectors, power heads, preverters, and then modeling and simulation of the combustion instability issue. We have got better computers now of higher speed. They can better model those things but it takes some investment and algorithms to try and get a better handle on that. Combustion instability is a phenomenon that occurs in less than a second, and you cannot stop it. So it will blow an engine up if it happens. The more you can do in computers, the less hardware you have to then have in your test program.

Senator UDALL. So you have got less than 1 second to get—

Mr. MITCHELL. Yes, sir, and you do not stop it. If it happens, it happens. You go get another engine.

Senator UDALL. Thank you for that insight.

Senator King?

Senator KING. Thank you, Mr. Chair.

I want to go back and try to push down a little bit more on this RD–180 decision. I guess the first question—and perhaps, General Shelton, you are the best person to ask. If you are not, perhaps one of you all can chime in. Ms. Kim, you may be. How serious is the interruption risk? Is this a theoretical risk, or is there any indication of an interruption in supply by the Russians? General Shelton?

General SHELTON. Senator, I will echo what Mr. Lightfoot said earlier. We have seen no indication of an interruption threat other than what Mr. Rogozin said. We have seen no indication from the commercial side. We have talked to ULA extensively. They have talked to their counterparts in Russia extensively, and there has been no indication that that is a serious threat at this time.

Senator KING. Now, even after Ukraine, Crimea, the various unpleasantness, no threats.

General SHELTON. Yes, sir. Certainly the potential is still there, but what we are seeing right now is business as usual.

Senator KING. And I want to press a little bit more on what it would mean. In answer to Senator Cruz's question, you have 15 of these motors in stock, in inventory in a sense. How many launches a year do we normally do? What do we have planned, say, for the next 5 years, total number of launches?

General SHELTON. We do roughly six or so a year of Atlas V, six, seven a year. So that is how many engines you are going to burn every year.

Senator KING. So basically we have a 2-year backlog of inventory.

General SHELTON. We do.

Senator KING. The other thing—and there may not be a short answer to this, but clearly one of the other things we have to ask about is the cost implications of developing our own engine. That is not going to be free.

General SHELTON. No, sir. And you have heard some projections here this morning, somewhere between $1 billion to $2 billion. The question then becomes can you stand not to pay that price or the potential of an interruption.

Senator KING. That is the question. My questions do not presuppose an outcome. I just want to be sure we are analyzing. This strikes me as a low-risk, high-consequence kind of situation. There is a low risk of this happening, but if it does happen, the consequences are high. Is that fair? Mr. Estevez, you nodded when I said that.

Mr. ESTEVEZ. That is fair. You know, again, the situation with Russia right now is volatile. So the risk is there. As General Shelton said, there is no indication that we would be cut off today. We can project into the future. So, yes, there is a good rationale for why we would move down the path to develop our own engine.

However, while we are doing that, use of the RD–180 is a cost-effective and proven way to launch our National security payloads.

Senator KING. Well, there was another factor, as I understand, in the late 1990s when the decision was made to go with the RD–180, other than the fact that it appears to be a high quality, reliable engine, is—it is kind of an odd thing to think about, but it was the desire to keep Russian rocket expertise in Russia. Are we not worried about that anymore, or is that no longer a factor? But that was apparently a national security consideration back when this decision was originally made.

General SHELTON. General Mitchell may want to comment more on this, but as I understand it, that was a consideration but certainly not the primary. If you look at this, this was really a commercial development as it started in the late 1990s. This was Lockheed-Martin building their own rocket, and they chose the RD–180 engine.

Mr. MITCHELL. I will just add to that that the conversations on the RD–180 actually started with General Dynamics before they were procured by Lockheed-Martin. The Russians came to the General Dynamics and said for $100,000 we can go modify the RD–171 engine which flies on the Zenet and produce you an engine that will be able to fit under your rocket. And it was a deal they could not pass up. So it was driven by a political situation but enabled by the cost benefit of doing it, and then the initial engines were only $10 million apiece. That cost has gone up, but initially it was very financially attractive to do it.

Senator KING. That is the kind of analysis that we have to do today. I mean, clearly in an ideal world, we would want to make our own engines and have control of that piece of the industrial base. On the other hand, this is a proven quality product, and there will be additional costs.

By the way, who makes this decision? Does Congress make this decision? Does the Air Force make it? Does the ULA? Who is going to decide when to move from the RD–180 to another engine?

General SHELTON. Sir, I would speculate that what would happen here is the executive branch would bring a proposal to the Congress and then the Congress ultimately has to decide whether or not to spend the money.

Senator KING. So you see it as part of the appropriations process in effect.

General SHELTON. Absolutely, sir.

Senator KING. I realize, Mr. Chairman, my time is expiring, but I would like to ask one more question.

I notice—a totally different subject—the purchase and the competition versus single source. Under the proposed rules for 2015 and 2016, there were going to be 14 competitive cores and 20 only ULA-capable. Under the President's budget for 2015, it is seven competitive cores and 20 ULA cores. It just strikes me—those 20 were inviolate? The competitive part got cut in half. The other got cut zero. Talk to me about that decision.

General SHELTON. It actually is a very involved answer. Many of those launches that were set aside for competition were GPS launches. As we looked at the health of the GPS constellation and we have decided that those are projected to live longer than expected, we did not need to procure the GPS launches on the schedule that we thought we needed to. So we have stretched that program out. That resulted in the loss of five of those seven that are no longer available for competition.

Another launch became too heavy, such that nobody but ULA could lift it.

And another launch was taken out for requirements reasons, and because we had a 36-core commitment to ULA as part of our pricing arrangement, we had to plug that hole that we had created by taking one of the requirements out. So that results in the seven.

So it really was not an anticompetitive thing, and as we said all along, it was up to 2014. And it is seven now. We think we may get an eighth in fiscal year 2015. But that is kind of where we stand, and that is the reason we have reduced the number available for competition.

Senator KING. Just as we were talking about the Russian engine creates risks, I think having a sole supplier creates risks for the country, not necessarily national security risks, but certainly financial risks. I believe that we need to move toward competition as rapidly and efficiently as possible just from the common sense competition is better than monopoly approach.

General SHELTON. And, yes, sir, we would absolutely agree with you. The advantage we have got with the current provider—it is a firm, fixed price arrangement. So we know exactly what the costs are.

Senator KING. Thank you.

Mr. ESTEVEZ. But if I could, Senator. The Department's position in working with the Air Force is to drive you that competition, and that is what we put in the program when we moved to both the block buy to decrease the cost and at the same time drove to competition. So, you know, the fact that the manifest moves around for budgetary and because of the health of the constellation reasons, it is not picking on those. It just happens to be that those are the ones—

Senator KING. You understand how it would appear.

Mr. ESTEVEZ. I absolutely do understand that. But I want to just reemphasize that we are committed to driving down that competition road to do this.

Senator KING. And I understand block buys are better than one off and you get a better price, and you have gotten that better price. But I think as a general principle, competition is where we ought to be heading.

Thank you.

Senator UDALL. Thank you, Senator King.

General Shelton, let me turn to you again. The EELV program mates its payload in a vertical configuration. Can you explain why that is done in terms of cost and risk?

General SHELTON. Yes, sir. We had to standardize how we were going to do this across our fleet. And it has to do with really fragile satellites, how they are manufactured, how the lifting mechanisms work, all that. So we standardize to vertical. So you basically take a payload out, encapsulate it in its payload faring. We lift it up vertically and set it down on top of the launch vehicle. And that has become our standard practice, and there are lots of good engineering reasons for doing it that way.

Senator UDALL. Let me turn to the Atlas V. It has a proven track record. Would you agree—and we also agreed that we want more competition, and Senator King's got to the specifics on the competitive side versus the block buy side. But would you agree that we need a tested and certified domestic alternative that meets all relevant performance criteria before we halt the use of the current engine?

General SHELTON. Senator, if we can continue to purchase RD–180s, that is the most economical approach. No doubt about it. If we got into a situation where that supply was interrupted and we had to go into some sort of crash program on development of an engine, that is a wholly different matter. My personal opinion, if we can continue to buy RD–180s, we ought to buy them. It is a good deal.

Senator UDALL. What resources are being utilized on the part of the Air Force to help SpaceX become certified for DOD launches? And when do we expect that process to be completed?

General SHELTON. Senator, we are spending 136 people on the problem and probably through fiscal year 2014, it will be somewhere around $60 million, probably approaching $100 million by the time we are done. And as I said earlier, if we can accomplish this on what we would consider to be a very green light schedule, they will be certified by December. As we look at what we are projecting with a higher confidence on the schedule, we think it is going to be the first quarter of fiscal year 2015—I am sorry—calendar year 2015.

Senator UDALL. I have got one last question. Let me restate my previous question. I am not sure you answered it in the way I was hoping—not the actual answer but just that you heard what I was asking.

Given the proven track record of Atlas V and the importance of competition in the launch market, would you agree that we need a tested and certified domestic alternative that meets all relevant performance criteria before we halt the use of the current engine? I think you said yes, but I want to make sure I was clear on that.

General SHELTON. I did say yes, Senator, because if you look at the manifest, Atlas V lifts about two-thirds of our manifest.

Senator UDALL. Thank you. I am going to yield to Senator Nelson.

Senator NELSON [presiding]. And I will still do cleanup.

Senator Cruz?

Senator CRUZ. Thank you, Mr. Chairman.

I would like to follow up on some of the very good questions that Senator King asked focusing on competition. General Shelton, how would you describe the benefits of competition in terms of acquiring engines and the capabilities for launch?

General SHELTON. Senator, I think there is no question that competition brings lower prices. It brings innovation and new ideas. What it cannot substitute for yet is reliability. We have a proven performer here, in fact, two lines, Delta and Atlas, that are very proven performers. So the question we have to answer is can we get to the place where we are as comfortable with a new entrant as we are with our current provider. That is why we have a very rigorous detailed certification process that is engineering-based, has 19 different engineering review boards that we will work our way through. That will have to be the substitute for numbers of launches. 72 in a row is a pretty good track record.

Senator CRUZ. When do you expect it is possible for all of these contracts to be competitively bid?

General SHELTON. Our schedule right now says that starting in fiscal year 2018, it will be a full and open competition.

Senator CRUZ. And between now and then, what is reasonable to expect?

General SHELTON. Between now and then, we have got the 36-core buy with United Launch Alliance. We will have at least seven, maybe eight launches available for competition, and there may be—who knows—some pop-up opportunities along the way as well. But we have a contractual agreement for a 36-core buy with ULA right now.

Senator CRUZ. And am I right in assuming that even with a competitive bid, it is entirely possible the current provider would win that bid?

General SHELTON. Absolutely.

Senator CRUZ. But the benefits of competition—the taxpayers may get a far more favorable price through vigorous competition than they would with a no-bid contract.

General SHELTON. Pencils will be sharpened. Yes, sir.

Senator CRUZ. Well, I would encourage expediting efforts down that road.

Mr. Estevez, you said a moment ago, if I heard you correctly, that what had been discussed here was a good rationale why we would go down the path towards development of a new engine. I want to understand that comment and reconcile it with the administration's statement of administration policy on June 17th where the administration objected to the House allocating funds to a new engine. Can you explain in your judgment what we should be doing towards developing a domestic engine so that we are not dependent upon Russian providers?

Mr. ESTEVEZ. There is a number of different paths that we can take to develop a new engine. What we said for that $220 million, I believe it was, is it is preliminary to be putting that money into the budget within the trade space of the budget at this point where we do not know the course that we are going to take to pursue development of a new engine.

Now, we have just asked for some reprogramming to do some risk reduction, and there is probably, as General Mitchell alluded to earlier, some time that you need to do that risk reduction before we decide whether it is going to be a public/private partnership that develops it, we will go to a commercial entity that will develop engines based our risk reduction, or it is inside the Government process to do that. So it is not that we do not want to go down the path in getting a new engine. It is the fact that the money was preliminary for where we are in that direction.

Senator CRUZ. What is your best case estimate for design, construction, test, and certification of a new engine both in terms of cost and time?

Mr. ESTEVEZ. Again, that will depend on the course that we select on getting to a new engine.

Senator CRUZ. But give me best case.

Mr. ESTEVEZ. 8 years.

Senator CRUZ. 8 years. So we talked earlier that if Russia cuts off these exports tomorrow, we do not have 8 years? worth of engines sitting in the warehouse. Is that right?

Mr. ESTEVEZ. That is correct.

Senator CRUZ. So if the ramp-up to develop a new engine is substantially longer than our capacity to survive not having these imports, it would seem there is some considerable exigency to starting that process now and not getting caught flat-footed if the worst comes to pass.

Mr. ESTEVEZ. Of course, if that happened, if we were cut off, we would use the stock that we have and we would allocate those in an interagency process. We would ramp up production of Delta, which can launch our manifest. As General Shelton said earlier, it would cause some significant delay and put some risk into our constellations, but we would do that. Commercial providers that we are pushing for our competitive environment will come on board. They will be able to launch some of the manifest. So there are mitigations. Now, those things will cost us money, and they will, as I said, put some risk into the time we get some of those constellations up.

In the meantimc, I would say throwing money at a problem that we do not know where we are going is not a good idea either at this point. So it is not just a matter of rushing money into a development of a new engine. We want to do that in a considered manner so we get the engine that we need.

Senator CRUZ. Although if you said it is 8 years, the longer we delay the beginning of that 8 years, the further out the end of that 8 years is.

Dr. Kim, do you have any thoughts on this question? We would welcome your thoughts as well.

Dr. YIM. So our study did not look to an independent schedule estimate assessment. So I cannot comment on that.

Senator CRUZ. Thank you very much.

Senator NELSON. And, of course, were the worst to happen, that it cut off today, in addition to the Delta heavy to launch the heavy payloads, assuming that SpaceX is certified by the end of the year, you would have that capability of launching medium-sized payloads. So all is not lost were he to do the unlikely thing of shut off Roscosmos.

Senator Sessions?

Senator SESSIONS. Thank you.

Well, I went past the Pentagon the other day, and we discussed once again—did they actually build this thing in 1 year? Mr. Estevez, this costs more—this delay. So here you have asked for $40 million—the Department of Defense has—for technology reduction for commercial new engine development this fiscal year through a reprogramming. It will be needed to develop strategies. We ought to have a strategy by now. How long does it take to develop a strategy? And initiate engine risk reduction efforts, technology maturation activities, early concept studies, and surveys. It goes on to focus on key risk components, technological development.

Well, I would just say if this was a private business and they got a major supplier that they no longer find reliable, they would get busy right now. So why can we not develop a situation now?

Now, I understand the Defense Department is predicting, Mr. Chairman, it might cost $2 billion to develop the engine. I have heard recently that one of the people who would be wanting supply it said they could do it for $840 million and would do it within 4 years and would put penalties on themselves if they did not produce that.

So, Mr. Estevez, do you think that is possible? Why do we not get busy? The fundamental question is are we going to continue with the Russian engine. Have we made a decision not to? And if we are not going to do so, which I think we have no choice but to make that decision, why do we not get busy and get this done and not drag it out?

Mr. ESTEVEZ. We agree that we should be moving away from the Russian engine. We want to use the Russian engine while it is available while we go through that development effort. Without sounding glib, it is rocket science and the development of new engine integration of that. So if there is a commercial company that is willing to go do that, we are happy to work with that, and that is one of the options that we are looking at is whether we can do this in the commercial sector, how much Government support is going to be needed. You have read our reprogramming action.

Senator SESSIONS. How many years do you project this to take?

Mr. ESTEVEZ. Our estimates are 5 to 8, Senator.

Senator SESSIONS. Five?

Mr. ESTEVEZ. 5 to 8 years.

Senator SESSIONS. Well, that is not acceptable.

Ms. Chaplain, do you think it is going to take 5—you mean to actually have the engine produced.

Mr. ESTEVEZ. Have the engine developed.

Senator SESSIONS. So how long will it take us to decide on what process we need, what kind of engine, and get moving on it?

Mr. ESTEVEZ. The Air Force is conducting an assessment right now that will be ready in the September time frame what we believe our best, most affordable course, affordable within the time frame course on development is. Again, that will look at public/private partnership, internal Government, or a commercial outsource on that.

Senator SESSIONS. Well, General Shelton, you know the history of this better than anyone. Do you agree that if we keep dragging this out, there is a danger we will slip back into uncertainty and delay, delay, costs go up, and maybe nothing ever gets done?

General SHELTON. That is a concern, Senator.

Senator SESSIONS. You have been at this for a long time. You are about to leave the Defense Department. Share with me what you think. Is there a danger and what do you think about it?

General SHELTON. We can stretch things out. We can make it longer.

Senator SESSIONS. Unwisely you mean.

General SHELTON. Absolutely.

Senator SESSIONS. Ms. Chaplain, do you have any thoughts? You have been around these programs for a long time yourself, and I think even some of your recommendations were taken into account, as you noted, saved $4 billion on the procurement that we have now, which was a good step. Do you have any idea how we could move quicker and less expensively in this crisis?

Ms. CHAPLAIN. I have been around long enough not to trust the numbers being thrown around today on either side and by vendors. But I do not believe——

Senator SESSIONS. You could put a penalty on a vendor.

Ms. CHAPLAIN. Yes, but sometimes you get stuck and when the problems happen, you go back to a cost-plus arrangement which the Government has done numerous times during the middle.

We do not know what we are actually pursuing right now. Is it a replacement, or is it going to be—what is the design going to be? And is it going to extend all the way into the design of a whole launch vehicle? The more extensive it gets, the longer it takes. And I agree that the need might be compelling, and if we lose time, we will be rushing activities even more later. And as you know, the more you compress and have to take on a lot of concurrency in your acquisition program to meet tight deadlines, the more you are at risk of having problems later on. So it is important to, first, figure out what it is we are really doing, get a good plan, and have disciplined processes in place. But I agree that if it is going to happen, we need to start working soon.

Senator SESSIONS. Well, thank you all. It was a good panel. It is an important issue.

General Mitchell, do you have any thoughts on this?

Mr. MITCHELL. Yes, sir. Just as one data point, when we were doing the EELV program, the RS–68 engine, which flies on the Delta, was a new development in a competitive environment. It took the contractor 6 years and $1 billion to develop that engine. That was without Government oversight at the time. That was straight commercial in the competitive environment where Boeing was competing with Lockheed-Marin, and it took them 6 years and $1 billion in 2013 dollars to do it. So that is a data point for you. And they were trying to be as aggressive as they could because they were in a competitive environment, and it was commercially developed.

Senator SESSIONS. Well, I just do not understand that.

Mr. MITCHELL. But that is a point for you.

Senator SESSIONS. My time is up.

But the plan was to develop an engine and many of the similar technologies in the Russian engine, nothing particularly new. And if we got busy on it, I think we would save money in the long run. The longer we delay it, the more alternatives we are going to have to use, more expensive launches, delaying of launches, and all that. I just wish we could go faster and make a decision.

Mr. Chairman, I am glad you are chairing this. Having flown on rockets and come back to be with us, it is not an issue that you do not know a lot about. You know a lot about it, and I appreciate you and Senator Udall for having the hearing.

Senator NELSON. My critics wish that I had gone on a one-way trip. [Laughter.]

Senator SESSIONS. I have to tell this story. I was debating. I thought my opponent had said something against NASA, and I said I think we should explore the solar system and go to Mars and I would like to go to Mars. And he jumped up and reached in his pocket and said I will be the first to contribute to sending you to Mars. I thought that was the highlight of his campaign. [Laughter.]

Senator NELSON. And he only got about 5 percent of the vote.

This is rocket science. So it is not easy and these decisions are not easy. And Ms. Chaplain has the historical perspective that somehow the cost of these programs grow, but they especially grow when you realize you are not just developing an engine. You are integrating it into a launch vehicle. You are going through a certification process, and then you have got to have the ground systems infrastructure. So does that add cost, Ms. Chaplain?

Ms. CHAPLAIN. Yes, it certainly does, and I do not believe that is all being brought into the mix at this point with the numbers we are hearing about.

Senator NELSON. What do you think about it, Mr. Dumbacher?

Mr. DUMBACHER. Well, I think, as you pointed out, Senator, this is a major, complex systems issue. It is not just an engine itself. We have technical issues we have to work out for a LOX/kerosene engine, but we also have to figure out how to integrate that into a launch vehicle. Typically when we develop engines, they are very integrally tied to the launch vehicle that we put them in. You cannot just move one engine from one launch vehicle to another very simply. And you have to go through the entire systems process, the ground systems, the logistics, and take into account all the complex technical interactions that you have to deal with in a design that is trying to go from 0 to 17,000 miles an hour in the space of a few minutes. It is a very complex systems approach. It has affect on the industrial base, and decisions that are made on one launch system affect other launch systems.

And it reaches across the Government. We know that if there is a major decision made with the NASA solid propulsion base, it affects the Navy's strategic D–5 missile program. We also know that if changes are made in terms of flight rate for the RS–68 and the Delta 4s, that has impact over on the RS–25 usage and costs for the space launch system. So we have to look at this access to space question from an overall systems perspective and account for all of the complexities that are in this, not just the engine itself. The engine is one key part of it, but there are larger other impacts that have to be addressed.

Senator NELSON. Plus, once you have got the new engine in the new rocket, then you have got to do payload integration, and that takes time and money in the new vehicle.

Senator King, you had a question.

Senator KING. Senator, you mentioned this was a record-breaking hearing, and it is for me because it is the longest Armed Services Committee hearing that I have been to where the word "sequester" has not been mentioned. [Laughter.]

And so I want to ask that question. I noticed the President's budget is listed as your planning budget, but the President's budget, as I recall, does not include the sequester. Does the sequester affect your procurement decisions, or are these forward procured, already contracted, and we do not worry about the sequester at least as far as these 36 cores are concerned, General?

General SHELTON. Senator, if sequestration comes back in 2016, which is, as you know, the law, we would have all kinds of priority decisions to make across the Department of Defense. What we have got right now are pricing agreements with ULA on that 36 cores. We do not have actual procurements. Those are done in that given fiscal year. So come fiscal year 2016, the buys that would be included in that fiscal year would be considered for whether or not those were priorities for the Department.

Senator KING. That is one of the charming effects of the sequester. It not only messes up budget planning, but it also could end up costing us money because we broke the 36 procurement block. We would end of paying more. Is that correct?

General SHELTON. Well, that would be a negotiation with United Launch Alliance, but I think the answer to that would be very likely yes.

Senator KING. So the short answer is that sequester would affect what we are talking about here today, whether it is planning for a new engine or the launches or the acquisition of various launch vehicles. So sequester is a factor in everything that we have been discussing. Is that correct?

General SHELTON. Yes, sir.

Senator KING. Thank you.

Senator NELSON. In other words, sequester is going to have an impact on assured U.S. access to space.

Let me do some cleanup here, and then you all jump in if you have any more.

Mr. Lightfoot, NASA is flying on SpaceX right now going to and from cargo to the station. All right. Now, General Shelton, you said that it is going to take an aggressive effort in order to get SpaceX certified to fly DOD missions. So what is the additional certification required to meet your certification needs that NASA has not already certified?

General SHELTON. And I will let Mr. Lightfoot talk about his side of this, but NASA has not certified SpaceX for their, for example, interplanetary missions. They are carrying cargo back and forth to the station, but in terms of the really high priority science missions, they have not certified SpaceX.

Mr. LIGHTFOOT. And I think that is an important point. Just to kind of frame it up here, there are classes of missions, and those classes range from A to D. This is simple stuff. But then there is a category of launch vehicles as well, and they are 1 to 3, 1 being the ones that we would take the most risk on, 3 being the ones that we would fly our most important payloads on.

Senator NELSON. Namely humans.

Mr. LIGHTFOOT. Yes, sir.

So what we have done is the missions that we are flying to the International Space Station with Orbital and SpaceX, the cargo is considered class D, which is the least level—where we are willing to take the most risk as an agency. So what we did is we did not do as much insight into that from a launch vehicle perspective as we would, say, on a class A mission.

However, what is really important is we focused most of our real oversight when they get close to the International Space Station because that is when critical activities can occur. So the activity associated with the birthing today of the Orbital Cygnus and the SpaceX Dragon—those we made them do a series of tests, a series of approaches, back-outs, all these things we do to ensure our own safety. So really what you are talking about is a risk categorization here in terms of the type of mission or spacecraft you are flying and the launch vehicle that it goes on.

For us, we are working on a Jason 3 flight that will be in December 2015. SpaceX is not certified for that flight yet. We are working through the certification process with them on that one, just like the Air Force is on the missions that they need. So we have them flying to the International Space Station, SpaceX and Orbital, but the certification for the next class of payload is the thing we are working on. And then there will be an even further certification, as you said, for when we have a commercial group provider as well.

Senator NELSON. Anybody want to add to that? [No response.]

Secretary Estevez, what about the cost of accelerating a new entrant certification compared to developing a new engine?

Mr. ESTEVEZ. Based on the costs of what we are doing now, obviously it is much cheaper to have a commercial provider, a capable, certified commercial provider, who can launch our payloads than it would be to develop a new engine. Now, right now we are not in a place where the providers that we have as new entrants can put up rockets that can launch the full manifest of payloads that are going on the Atlas 5.

Senator NELSON. You could with the Delta 4.

Mr. ESTEVEZ. I could with the Delta 4.

Senator NELSON. But that is going to cost some more.

Mr. ESTEVEZ. Yes, absolutely true. Development a new engine and the integration costs of that are obviously much more expensive than the cost that it costs us to certify the new entrant. But, again, I want to make sure that we are certifying new entrants that are capable of launching the payloads that we are launching. Otherwise, I will be sitting up here and General Shelton's successor about why we launched into the ocean, and I do not want to be doing that either.

Senator NELSON. Mr. Dumbacher, have we missed any other potential options other than a new launch vehicle or engine development that could address this RD–180 potential problem?

Mr. DUMBACHER. I think you have seen, Senator, across the board from the Department of Defense and NASA the involvement with General Mitchell's study and what it would take to replace the RD–180. That was all good work and I do not need to refute or take on any of that.

I think, again, back to your previous question, I think my caution would be that we make sure we address this from an overall systems perspective and a larger perspective than just an engine replacement because it does have ramifications to other launch systems. And these decisions are long-term and have ramifications for lots of years. The decision that this country made at the end of the Apollo program to dramatically reduce our work on LOX/hydrocarbon engines is still playing out today and is part of this conversation that we are having this morning. So I think we need to be aware of that, that these decisions are long-ranging, have large impacts, and the unintended consequences that can be had with any of these decisions we need to think carefully through to make sure that we do not inadvertently end up in a place we do not want to be as a country.

Senator NELSON. Just to make it more complicated, for example, you have already mentioned the impact on the Navy's rockets, which are solid rockets. If you do not have a solid rocket program in the other departments, that means the cost to the Navy is going to go up.

Mr. DUMBACHER. That is correct. In the past, NASA has worked with the Department of the Navy in the strategic missile program on what the impacts would be to them from an industrial base perspective and a supply chain perspective if NASA were to do something different than the solids. We understand that. NASA understands what that industrial base implication is, and we have to be wary of that.

We also, as I mentioned earlier, need to be aware of the ramifications and the impacts on the liquid booster side between RS–68 and the supply chain that is shared between the RS–68 and the RS–25.

In the end, I think our problem has been—in my view, the issue is getting the cost down and what we need to do to get the cost down. This Nation has spent over the last 40 years making significant investments in LOX/hydrogen, solid propulsion expertise. We are the world leaders in that, and I think it behooves us to look at those possible solutions as part of the overall system implications.

Senator NELSON. Can anybody on the panel give us the historical perspective of how many programs we have actually started and then canceled and how this plays into this decision? How can we ensure that if we start this new program which, as I mentioned at the outset, some of us on the Armed Services Committee have put $100 million in this coming fiscal year to start it—how can we be assured that this is not going to get canceled in a few years and therefore the waste of the money?

Mr. ESTEVEZ. I cannot give you the number, Senator, of how many programs—I am sure we could get you that—over the course of time. From an acquisition perspective, one of the things that we are trying to accomplish under Better Buying Power is do not start programs unless you are going to fund them and you are going to put the right structure in place to follow through on those programs. On the development side, there is always going to be some growth, especially in a high risk program like this that is complex. But we would have to commit the dollars in the budget. And again, that goes back to my earlier things of how are we going to do this and we are not sure what the course would be to develop a replacement whether it is commercial sector or public/private partnership and the whole integration of that for the RD–180.

I will go to Senator King's point on sequestration. Of course, that also impacts the point on the budget and where that trade space is related to this.

Senator NELSON. Senator McCain?

Senator MCCAIN. Thank you, Mr. Chairman. I thank the witnesses.

General Shelton, you are widely quoted as saying generally the person you are going to do business with you do not sue. Do you stand by that statement?

General SHELTON. Senator, the context for that was the conversation on the litigation between SpaceX and the Air Force, and yes, I do stand by that statement. We are trying very hard to get them certified and spending a lot of money, a lot of people.

Senator MCCAIN. First of all, what about the fact that already there is a suit pending by the ULA subsidiary seeking $400 million in additional payments from the Air Force? In other words, if some company or corporation thinks that they are not being fairly treated, you do not think that they should be able to sue? I mean, that is not our system of Government, General Shelton. I do not really get your statement except that it shows real bias against the ability of any company or corporation in America to do what they think is best for their company or corporation. And a subsidiary of ULA is suing for $400 million. Do you think they should be suing?

General SHELTON. Senator, that is over a technical payment situation.

Senator MCCAIN. Oh, I see. So it is okay it is over a technical payment situation but not any other. General Shelton, you have really diminished your stature with this committee when you decide whether people or organizations or companies should be able to sue or not and make comments about them.

Ms. Chaplain, it seemed all of a sudden that the Air Force now found out that GPS satellites would now be able to stay up longer. Was that not known for a long time?

Ms. CHAPLAIN. They do analyses of the constellation on a regular basis and see how they are going to last. They tend to make adjustments to the manifest based on that. It is just not unusual to see changes, though the ones that were made this year were a little more substantial than usual.

Senator MCCAIN. So the decision to cut the competitive launches even more by delaying launches really should not have come as a surprise.

Ms. CHAPLAIN. It is never a surprise to me generically that there are changes to the manifest either based on budgetary reasons or the length of a constellation. I have never believed you should trust what that manifest is year to year.

Senator MCCAIN. Facts are stubborn things.

Mr. Kendall, who we had extensive conversations with when he came up in the Armed Services Committee for his job because of the failure—is it not true that ULA has breached Nunn-McCurdy more than once or twice? Is that true, Ms. Chaplain?

Ms. CHAPLAIN. At least the last one that I know of was the most recent one. There may have been one before that.

Senator MCCAIN. So at least we know of one breach of Nunn-McCurdy, which is cost overruns of a dramatic and significant amount. That did not seem to bother anybody in the Air Force or the industrial complex because now, instead of increasing the numbers of competitive launches, we have decreased the competitive launches to an outfit that breached Nunn-McCurdy because of cost overruns. How does that give them any credibility? Do you want to respond to that? You do not have to.

So now, Mr. Estevez, we will now see a total of three, although perhaps the Congress will mandate at least one additional competitive launch, and that is fine with you. Is that right?

Mr. ESTEVEZ. Senator, we are committed to the competitive course. We are aggressively pursuing to get SpaceX certified so that they can launch our satellites. They do not have the capability based on their current certification process to launch the full manifest of those satellites, but we look forward to getting them to be able to launch the ones that we have up for competitive—that they are capable of launching.

Senator MCCAIN. So even though they have just completed a third successful launch. Is that right?

Mr. ESTEVEZ. That does not complete the certification process, Senator.

Senator MCCAIN. I know that. But the certification was supposed to take place in January. Is that correct?

Mr. ESTEVEZ. They are on their path to certification. It was not supposed to be completely certified in January to my knowledge.

Senator MCCAIN. And do you know when you will make a final decision?

Mr. ESTEVEZ. As General Shelton discussed earlier, if everything goes well with their certification process, they should be certified by the end of this year.

Senator MCCAIN. Mr. Lightfoot, NASA introduced launch competition into its processes by having two competing companies for the commercial audit transportation service contract. Have there been benefits of that, Mr. Lightfoot?

Mr. LIGHTFOOT. Yes, sir, we think we have gotten a good value in the process. The payloads we are launching are what we call class D because we have the two providers to get to the International Space Station.

Senator MCCAIN. Well, let us talk about the so-called—this great savings that is supposed to take place with a block. You saved—quote, saved—you are arguing, General Shelton, that the Air Force repeatedly said it has saved $4.4 billion on space launch costs by awarding a sole source block buy contract to ULA, disregarding the fact that ULA breached Nunn-McCurdy, which required the notification to Congress of cost overruns. But it is really cost avoidance.

Ms. Chaplain, do you have a view on that of whether that is actually a, quote, saving of the $4 billion which was advertised because of the sole source block buy contract to ULA?

Ms. CHAPLAIN. What it represents is—the ultimate price that they negotiated was substantially lower than the price they started out with in the negotiations. We did not investigate the exact $4.4 billion and what was behind it, but we do know that the Air Force took a number of actions to arm themselves with better information for the negotiation process, principally getting more information on costs and pricing in preparation.

Senator MCCAIN. Well, actually the Office of Management and Budget refers to cost savings as a reduction in actual expenditures. That has not occurred in the EELV program.

I think, Mr. Chairman, that the issue of Russian rockets has been already pretty well massaged, but the fact is we are seeing here—and I do not predict, but a few years ago, there was a situation concerning the Air Force tanker. And I did not like it at the time and I fought against it at the time. And people went to jail and people were fired. I do not like this deal. I do not like the fact that we are now going to have basically maybe three—or if the Congress has its way, four—competitive space launches given to an outfit that has breached the cost overruns to the degree that it required notification to the Congress of the United States of cost overruns.

I thank you, Mr. Chairman.

Senator NELSON. Thank you, Senator McCain.

When you look at the value of competition, it is clearly well established. For example, NASA is going through a competition now of human rating rockets to take astronauts to and from the Space Station, and there are probably at least three competitors in that competition.

Senator MCCAIN. And what we are seeing here, Mr. Chairman, is a reduction in planned competition for whatever reasons. The actual reality is, despite Mr. Kendall's admonition to increase competition, we are seeing a decrease in competition. And then when the company does not like it and goes to court, they are criticized by a uniformed officer who really has no business talking about the conduct of a corporation as to what their legal options are.

Senator NELSON. Well, the Russians have certainly brought this to a head. And at the outset of the hearing, it was mentioned that it was the policy of the U.S. Government back in the 1990s, once we decided to buy the RD–180 from the Russians because it is an excellent engine and it was cheaper and we were employing Russian engineers and scientists instead of them going elsewhere on the planet—but it was the policy of the Government at the time we were going to develop a follow-on engine. That got put aside. So we are where we are particularly because of the deputy prime minister's sarcastic comments from Russia even though the statements were said—he made them at the time—he was only going to not supply the RD–180 for military launches. He is still going to provide them for civilian. But you will notice there was not a peep out of Roscosmos. They obviously want to continue that.

But, nevertheless, it brings it to a head, and it brings us to the table today. And these are complicated decisions, multifaceted, involving many different programs, but all of which come down to the bottom line, assured access to space for the United States.

We want to thank you all. You have been most enlightening.

And the meeting is adjourned.

[Whereupon, at 11:43 a.m., the subcommittee adjourned.]

www.ingramcontent.com/pod-product-compliance
Lightning Source LLC
Chambersburg PA
CBHW080621180526
45168CB00007B/3008